# Meaningful Complexity

*De Leon Galileo*

De Leon Galileo

ISBN: 978-1-961028-91-3

# Dedication

To the loving memory of Romeo and Pacita, my parents Giovanni and Gerardino, my brothers Julie Widjaja; Rosa DeLeon; Ramon, Felicitas, and Regie Bagasan; Cely Pornobi Sonny Balak; Alex Tan.

# Acknowledgment

The families of: DeLeon; Wu; Widjaja; Tan; Bersamira; Rarela; Cabatbat; Balak; Vu; DeDios; Silwany; Rahman; Ramos;Guerero; Gutierrez; Jean-Francois; Flores; Punsalan; Burns; Bulan; Mahon; Yoo; Pornobi; Bagasan; Panganiban; Carreon; Gellman; Miranda; Valbuena; Amazon Publishing Team: Emma Baker; Mark Anderson.

# Table of Contents

# About the Author

De Leon Galileo was born in the Philippines. In 1970, his parents and the family immigrated to the United States in New York. He is a self-proclaimed child psychologist, having babysat 4 kids from his brothers and sisters over several years. During this time, he found little use for his associate degree in engineering.

Currently, he is serving babysitting duties to his grade school grandnephew and taking over the responsibility of feeding 3 beautiful stray cats from a neighborhood friend who recently passed away.

# Chapter 1: Introduction

Snowflakes, Hurricanes, and Cyclones are orders that nature can extract from chaos. However, the claim that life was also extracted out of chaos creates a disconnect between the observable formation of cyclones and the unobservable formation of life.

1. Either, there might be another type of order-extracting mechanism.
2. Or are order-extracting mechanisms traceable to some abilities?

There are differences in complexities between the extraction of order that produces cyclones and the order that creates life. The disconnect is in not realizing that there are two types of order-extracting mechanisms. Scientists are treating the two mechanisms as the same.

This is similar to when two fractions with different denominators are added.

We exist in the physical reality, but at the same time, we experience the non-physical reality of the non-physical activities of our minds. Materialists are mainly focusing on physical activities in physical reality. At the same time, ignoring the non-physical activities of the unobservable non-physical reality of the mind.

How do we know that there are two types of realities?

Under a microscope, scientists can verify that a brain is made of matter; however, that same microscope cannot tell scientists what messages are in a person's thoughts, nor if those messages have height, width, length, and weight. Information or messages in the mind cannot be read physically; they must be inferred mentally or expressed in some physical medium.

Denying the existence of the non-physical reality of the mind does not remove the reality of the experience of us having an awareness of it.

Therefore, there must be at least two types of realities:

1. Material State
2. Non-material State

For an extraction of some order to happen, some ability would be required. If there are two types of realities, then there must also be two types of abilities:

1. Physical
2. Mental

# Chapter 2: Nature's Physical Abilities

"There is in this Universe much of what seems to be designed. But instead, we repeatedly discover that natural processes, collisional selection of worlds... can extract order out of chaos, and deceive us into deducing purpose where there is none"

*Carl Sagan*

Two possible sources where abilities to extract order could come from:

1. Natural (nature)
2. Supernatural (God, Intelligence, Spaghetti Monster)

If abilities come from Nature, then what is Nature? Nature has no physical body, nor can it be found in some specific address. It is simply a concept represented by the 4 fundamental forces and the chemical bonding of particles. According to many scientists, these four are the main mechanisms behind creating all that exists in the Universe.

If abilities come from a supernatural Being (God or Intelligence). What, then, is God or Intelligence? The belief in the existence of God is simply an assumption. Like Nature, God or Intelligence has no physical body, nor can it be found in some specific places.

The 4 fundamental forces cannot be assigned to the Supernatural Being because scientists have already assigned those forces to Nature.

What can be assigned to God or Intelligence are the 4 foundational mental abilities based on human intelligence.

Here, we are assuming that these four mental abilities are similar but not equal to God's mental abilities. An assumption that is no different from the scientists' assumption that the 4 fundamental forces were created by the Big Bang.

The difference between Nature and God is that the concept of Nature is the concept of Materialism, basically the concept of Atheism. On the one hand, Materialism is based on the lack of belief in God but a belief in Nature and the creative mechanism of chance. On the other hand, Creationism is the lack of belief in Nature but a belief in God and the creative mechanism of Intelligence.

The claim that Materialism or Atheism, which is the lack of a belief in God, makes Materialism or Atheism, not a belief system, cannot be true. If Creationism is a belief system, then Materialism is also a belief system. Because word pattern for word pattern, the two are similar.

Nature has no mental abilities but has 5 physical abilities. These are:

4 Fundamental Forces and the Chemical Bonding of Particles.

**Gravity,** "...gravitational attraction between two bodies increases in proportion with their masses, diminishes with the square of the distance between them..."

*Richard Welob*

**Weak Force,** "...Original d-quark is changed to a U-quark..."

*webs.morningside.edu/...*

**Strong Nuclear Force** "... hold a nucleus together against... repulsion (electromagnetic force) of the protons..."

*Nick Connor (2019, 05:22)*

**Electromagnetic Force,** "... positively charged atomic nuclei to attract negatively charged electrons allowing atoms and molecules to form ..."

*ecuip.lib-uchicago.edu/...*

**Chemical Bond,** "... a weak or strong electrical attraction that holds atoms ... two or more atoms held together by chemical bonds is ... a molecule ..."

*Lindsay M. Biga, Sierra Dawson*

- Supernatural Being, Physical Abilities (not known)
- Human's Physical Abilities (hands, legs, bodies)
- Supernatural Being, Mental Abilities (4 foundational mental abilities - assumed)
- Human's Mental Abilities (4 foundational mental abilities - covered in Intelligence Section)

From the two types of sources (Nature and God) where abilities could come from, two types of Evolution can occur:

1. Material-based Evolution or Property-based, order-extracting mechanism or Accidentally-Arranged Formation of Matter
2. Intelligence-based Evolution or Intelligence-based, Order-extracting mechanism or Intentionally-Arranged Formation of Thoughts and Matter

Material-based evolution is a change in the structure of matter caused by natural forces with a single-ability function dictated by its property. And whose interactions are dependent on the accidental motion of matter.

Examples of Material-based Evolution:

- Planets, Stars, Galaxies, Elements, and Snowflakes
- Plate Tectonic, Earthquake, Hurricane, Tornado, Blizzard, Tsunami, Forest Fire, Sandstorm

Property of matter is the main order-extracting mechanism of Nature.

Intelligence-based Evolution is a change in the non-material thoughts, physical objects, and physical anatomy caused by the 4 foundational mental abilities of intelligence (based on human intelligence). The physical interactions are intentional and objective-based mental activities first, then applied or converted into physical activities second.

In addition, Intelligence-based evolutions are higher forms of order extractions. And have the characteristics of proportionality, symmetry, uniformity, and specificity. In other words, intelligence is the main order-extracting mechanism of the assumed Creative Being as well as by human beings.

An example of an Intelligence-based evolution is "knowledge," which grows or improves as more information is added or learned about the subject. There are different fields of knowledge: Language, Music, Math, Psychology, Engineering, Biology, and Auto Mechanics.

Examples of Intelligence-based evolution on anatomy (based on the unknown assumed Creative Being):

A. Embryo to fully grown organism, animal, or human
B. Micro Evolution - Small changes in species
C. Macro Evolution - Big changes in species, such as Dinosaurs to Birds, are not observable. (scientists' assumption)

Examples of Intelligence-based evolution on physical objects:

A. Projectiles - Stones to Spears, to Arrows, to Cannons, to Missiles
B. Communication - Hand signals to Smoke signals to Flag signals to Mirror signals to Telegraphs to Wired Telephones to Cell Phones

Something must go through some process for it to evolve or become in order. Any process must have a starting point where the formation, selection, or becoming orderly begins. And an ending point where the process stops once the formation, selection, or order is complete. To have what can be considered a process requires the ability to do a set of functions:

1. Starting
2. Stopping
3. Continuing
4. Ending
5. Repeating

And to be able to do higher forms of order extractions or complex order-extracting activities requires mental and physical abilities (based on human mental and physical abilities).

"... 11-year-old ... Isabella Brazhnikova ... from Auckland, New Zealand ... Her hyper-realistic drawings of animals captured the hearts of viewers from all the nations ..." *Isabellaclever.com*

Isabella Brazhnikova exemplifies an intelligent agent who can intentionally arrange patterns of colors or intentionally extract higher forms of meaningful, complex, colorful patterns of order.

Nature, however, because it lacks intelligence, it cannot therefore have:

A. The mental ability to identify planets, size, distance, objective, coding, Laws of Physics, or Hyper-realistic drawings as information.
B. The mental ability to attach meaning to specific items to form knowledge.
C. The mental ability to store items of knowledge as items of information in a memory it does not have.
D. The mental ability to understand a specific item of knowledge it never attached meaning to.
E. The mental ability to formulate a response to any specific intervening information it cannot mentally understand or identify as existing information.

Nature is then left depending mainly on its physical abilities, which comprise the 4 fundamental forces and the chemical bonding particles. However, these 4 fundamental forces are each dependent on a single property. Each property gives each force the ability to do only one physical function as its main order-extracting mechanism.

Unfortunately, having the ability to do only one physical function does not allow each force to do multiple physical functions or complex physical order-extracting activities. Even when the 4 fundamental forces are combined with other natural forces such as earthquakes, hurricanes, tornadoes, blizzards, tsunamis, forest fires, and sandstorms, it would still not be enough abilities to take any material-based evolved order into the higher stages of order, or meaningful, complex orders.

Examples of complex, material-based evolved orders are:

1. Pile of Snow, Pile of Rocks, Pile of Alphabets, Pile of Color Pencils.

a. A pile of Snow, Rocks, Alphabets, or Color Pencils is complex but not meaningful. It would be meaningful if the 4 fundamental forces could turn a pile of material items into something useful, such as:

   i. Building Homes for the Homeless

   ii. Building Bridges for the Migrating Animals Crossing the Rivers in Africa

   iii. Dismantling all Nuclear Weapons

   iv. Reconfiguring all Military Tanks into High-Speed Trains for all Nations to promote Peace

For each of the 4 fundamental forces to be able to reach the higher stages of order extractions or meaningful complex extractions without using intelligence. Each force would need four additional properties:

1. The property to physically organize material items
2. The property to physically control material items
3. The property to physically manipulate material items
4. The property to apply time to each function of
   a. Starting
   b. Stopping
   c. Continuing
   d. Ending
   e. Repeating

It is not an ability where the 4 fundamental forces and chemical bonding of particles cannot start, stop, continue, end, or repeat any process of extracting order at a chosen time.

# Chapter 3: Applicability of Nature's Abilities

If the 4 fundamental forces and chemical bonding of particles were capable of getting to higher stages of order extractions or meaningful complex order extraction processes, then there should exist planets in the shape of a hammer, a piano, or a Statue of Liberty.

All these higher stages of order extractions, or complexly extracted planetary shapes, should easily be verifiable by the James Webb Space Telescope.

In the early 1970s, Carl Sagan had the opportunity to demonstrate the ability of the 4 fundamental forces and chemical bonding of particles to extract order out of chaos.

"... But, instead, we repeatedly discover that natural processes, collisional selection of worlds ... can extract order out of chaos ..." *Carl Sagan.*

He and his team of scientists had the vision of sending a probe into deep space. Their objectives were to take pictures of Jupiter and other objects that the space probe could encounter beyond our Solar System. He could have tested his claim by spreading all the space probe parts on the floor of an empty large room in NASA's Jet Propulsion Lab. Those parts would have included parts from a washing machine and a refrigerator.

The objectives would have been:

1. For Nature to extract order out of the chaos of parts that are spread on the floor by assembling a space probe using only the 4 fundamental forces: Gravity, Weak Force, Strong

Nuclear Force, Electromagnetic Force, and Chemical Bonding of Particles

2. To measure the speed with which the 4 fundamental forces and chemical bonding of particles are able to extract order or assemble the space probe

3. To observe the physical abilities of the 4 fundamental forces and the chemical bonding of particles to:

   a. Physically extract or organize parts of the space probe by separating the parts belonging to the space probe, and the parts belonging to the washing machine and refrigerator

   b. Physically control parts so that only the space probe parts are being used in assembling the space probe

   c. Physically manipulate space probe parts so that each part is attached to its proper connective places

   d. Apply time to each function of Starting, Stopping, Continuing, Ending, and Repeating any order-extracting process

Carl Sagan surely must have relied on the 4 fundamental forces and chemical bonding of particles to extract order or assemble the space probe naturally. After all, the ability of Nature to extract order out of chaos was what Carl Sagan had been promoting most of his adult life. He could not possibly have been promoting something he had no faith in applying.

Instead, Carl Sagan relied on other teams of scientists, engineers, and technicians, who all applied the principles of human thinking and the 4 foundational mental abilities of human intelligence to extract order or build Voyager I.

# Chapter 4: Natural Selection

"Step-by-step, natural selection could drive this transformation to increased complexity because each intermediate form would provide an advantage over what came before ." *Scientific American.*

Natural Selection is a specific type of Activity in Nature. An example is a case study on Peppered Moths. The study describes how during the Industrial Revolution in England, pollution had darkened the barks of many local trees, affecting the moths living in those areas.

The light-colored peppered moths who rested on those trees that had darkened from pollution became an easy target for birds on the hunt for food.

After a period, the dark-colored peppered moths, which blended well against the darkened trees,  managed to evade the hunting birds, allowing the surviving dark-colored peppered moths to pass their genes to the next generation of dark-colored peppered moths. Thus, resulting in an increase in the dark-colored peppered moths' population. At the same time, the light-colored peppered moths had the opposite effect.

Although the 4 fundamental forces were present during the bird and moth interactions, the forces were not active participants.

- Gravity did not pull the birds toward the moths.
- Strong Nuclear Force did not make the dark-colored peppered moths undetectable to the birds.
- Weak Nuclear Force did not weaken the birds' perception of the dark-colored peppered moths as food.
- Electromagnetic Force did not cause the increase in population among dark-colored peppered moths.

- Chemical Bonding may have darkened the trees, but it did not cause the moths to rest on those trees.

Therefore, underneath the seemingly natural consequences of events lie the bird's intelligence as the true mechanism of change.

Intelligence allowed the birds to identify the light-colored peppered moths resting against the darkened trees. The darkening of the trees was simply a circumstantial convenience to the dark-colored peppered moths.

Ignoring the ability of the bird's intelligence to identify food and just focusing on the consequences of events and then naming those consequences of events as Natural Selection takes away the role of the bird's intelligence as the main Mechanism of Change. It instead gives that role to Nature.

By giving Nature the title, The Mechanism of Change, scientists have given nature an imaginary ability that gives the appearance of Nature having the ability to do something it really has no ability of doing. For example, Nature causing a change in the dark-colored peppered moths' population without directly participating in the bird and moth interactions.

The words "Natural Selection" mislead people into believing that all animal interactions are natural events that can happen without guidance from any outside source. If the birds did not have intelligence and could manage to catch only light-colored peppered moths, or moths are just naturally falling into birds' mouths, then scientists would be justified in referring to these consequences of events as Natural Selection.

Natural Selection, Sanctions, Stimulus packages, and Bailouts have the same effect as Selective Breeding of Animals. Each is selecting some feature, item, or entity either to be nurtured, oppressed, or destroyed. All selective processes are, therefore, intelligence-based evolution, intelligence-based extracted order, or intentionally arranged order because intelligence is behind every selection process doing the selecting.

"... step-by-step, natural selection could drive this transformation to increased complexity because each intermediate form could provide an advantage over what came before ..." *Scientific American.*

The idea of a step-by-step process moving toward higher complexity is really the same as the Beneficial Trait argument, where the continuous addition of beneficial traits would lead to higher and higher complexity. But why would matter, protein or a simple organism, hold on to some feature that is beneficial to itself?

For matter, protein, or a simple organism to hold on to some beneficial feature, suggests that it is aware that some information or force is causing its condition to worsen. Therefore, it is reacting to its worsening condition by holding onto some features it finds beneficial to itself. Perhaps the feature it is holding on to has the ability to extend its existence.

However, if matter, protein, or a simple organism has no intelligence, then it cannot have the mental ability to identify conditions, beneficial traits, existence, deterioration, and self as information. It does not have the mental ability to attach meanings to items of information to form knowledge. It does not have the mental ability to store these items of knowledge as items of information in a memory it does not have.

It cannot have the mental ability to understand items of knowledge it did not attach meaning to. Therefore, matter, protein, or a simple organism cannot just suddenly perform a function of holding onto a feature whose characteristics it has no understanding of. If matter has no mental awareness of its existence, it cannot obviously be aware that it is holding onto anything, beneficial or not.

# Chapter 5: Genetic Coding

Microbes, plants, fish, amphibians, reptiles, birds, apes, and humans all have DNA in common. What is DNA?

". . . DNA or deoxyribonucleic acid, is the hereditary material in humans and almost all other organisms . . . The information in DNA is stored as a code made up of four chemical bases: Adenine ( A ), Guanine ( G ), Cytosine (C ), and Thymine ( T ) . . ." *National Library of Medicine*

DNA operates on the mechanism of a code. However, genetic coding is not a property of matter but an activity of the mind.

DNA is perhaps the highest form of complexity. It is not the number of parts that makes the DNA complex but the intelligence behind the intentional arrangement of parts that makes its complexity meaningful.

DNA encodes for proteins which are made of amino acids. And amino acids are made of carbon, hydrogen, nitrogen, and oxygen.

Proteins are therefore made of matter. DNA must therefore be encoded only for things that are material. Such as eyes, skin, muscle tissues, and bones. Or the measurement, proportionality, symmetry, and uniformity of human and animal anatomy.

DNA, however, does not encode for non-material mental activities, such as:

1. Our Mental ability to identify and understand information
2. Our mental ability to formulate objectives. Such as:
   a. World Peace
   b. Nuclear Disarmament
   c. Lower Interest rates for the poor

    d. Affordable housing

    e. Feed the Hungry

1. Emotions:
   a. Love
   b. Forgiveness
   c. Aggression
      i. War
      ii. Intimidation
      iii. Manipulation
      iv. Suppression
      v. Rape
      vi. Wife- Beating
      vii. Robbery
      viii. Murder

2. Conceptual Ideas:
   a. Cell Phones
   b. Computers
   c. High-Speed Trains
   d. Electric Cars
   e. Systems
   f. Instructions
   g. Laws of Physics
   h. Hyper Realistic Drawings

3. Mental Disorder:
   a. Narcissistic
   b. Paranoia
   c. Bipolar
   d. Schizophrenia

4. Ideologies or Philosophies :
   a. Religious
   b. Political
   c. Military
   d. Science or Institutional

5. Knowledge:
    a. Math
    b. Science
    c. Language
    d. Arts
    e. Sports
    f. Engineering
    g. Biology
    h. Psychology
    i. Artificial Intelligence
    j. Auto Mechanics
6. Styles / Preference
    a. Hair
    b. Clothing
    c. Music
    d. Architectural

If there exists a code, then there must also exist a mind, which has the mental ability to formulate or arrange messages in the form of instructions for creating proteins and the mental ability to identify which relevant messages must be included in a code, and arranged in a grouping pattern of nucleotides. A genetic code is a communication tool designed to efficiently carry messages toward a specific end, destination, goal, or objective, such as providing proteins to humans and animals whose existence depends on those proteins.

# Chapter 6: System

What is a system?

"…System is a set of rules, or arrangement of things, or a group of related things that work toward a common goal." - *yourdictionary.com/ system*

Creating a system requires 3 things:

(higher levels of-order forming systems)

1. **Intelligence** that identifies:
   a. Activities - which are relevant to the objective. For example, creating proteins.
   b. Working Mechanisms - which are relevant to the objective.
   c. Objective
2. **Working Mechanisms**, for example:
   a. Genetic Strand
   b. Nucleotides
   c. Enzymes
   d. Ribosomes
   e. Cytoplasm

Working mechanisms will produce only proteins as their designated product, which is relevant to the objective.

3. **Objective** - To service some entity that would have a need for proteins.
   a. For example, Humans or Animals

A living Cell which is part of the autonomic nervous system, is a system in itself because it is made of a group of independently working components of DNA, RNA, Enzymes, Cytoplasm, and Ribosome, all working as a unified mechanism. Each independently working member component contributes a product or a service toward achieving the specific objective of creating proteins.

# Chapter 7: Automation

An important feature built into the DNA as well as in all body organ systems is automation. A feature that is often taken for granted.

An automated body organ like the heart gives us a maintenance-free body mechanism that allows us to participate in other activities without having to worry about manually pumping blood and oxygen into our cells on a 24-hour basis just to exist.

Automation is not something that can be assembled by random bonding of particles. Since matter has no intelligence, it cannot have the mental ability to identify what automation is. Nor have the mental ability to understand what automation does. Automation is, again, not a property of matter but an activity of the mind.

# Chapter 8: Fine-Tuning

What is Cosmological fine-tuning?

"Consider the expansion rate of the Universe . . . A change in its value by a mere 1 part in 10 to the 120th parts would cause the Universe to expand too rapidly or too slowly. In either case, the Universe would again be life prohibiting…" – *drcraigvideos 2017*

Fine-tuning, like Genetic Coding, is not a property of matter but an activity of the mind. A stable cosmological environment that suits human and animal existence can occur from the fine-tuning of the Universe could not have been a mere coincidence. Considering that proteins being specifically created by DNA to keep humans and animals alive would have been pointless if a stable cosmological environment was not available.

Could there be other Universes out there? And could our Universe just happen to have a gravitational strength value that suits our existence? A multiverse would mean each Universe would have a cosmological environment with a specific gravitational strength value.

These differences in gravitational strength value would range from 1 in 10 to 1 power up to perhaps 1 millionth power. Depending on the number of Universes within a multiverse.

The problem with a multiverse is that each Universe would need to be maintained. However, the ability to maintain any specific gravitational strength values in an incrementally increasing order would require intelligence. Because only intelligence would have the mental ability to:

1. Identify different gravitational strength levels.
2. Extract, create, or arrange the order in the form of different gravitational strength levels.

3. Apply specific gravitational strength levels to each of the Universe within the multiverse.

4. Maintain gravitational strength at a specific level for each Universe within the multiverse.

5. Calibrate gravity by incrementally increasing strength levels so as to avoid multiple Universes all having the same Gravitational strength level.

However, Nature has no intelligence and therefore has no mental ability to conceptualize or identify Value, Maintain, Gravitational Strength, Universe, Order, or Increment as Information. It does not have the mental ability to attach meaning to such items of information to form knowledge. It does not have the mental ability to store items of knowledge as items of information in memory it does not have.

It does not have the mental ability to understand items of knowledge it never attached meaning to. It cannot then perform the function of maintaining gravitational strength level without understanding how gravity could be maintained at a specific strength level. Or that having different gravitational strength levels could affect the environmental condition of each Universe. Gravity would simply jump from one gravitational strength level to the next so that it could never form a multiverse, let alone a Universe.

Our finely-tuned Universe has 3 components that qualify it as a system:

1. **Intelligence** - which identifies:
   a. Relevant Activity - such as calibrating gravitational strength
   b. Relevant Working Mechanisms
   c. Objective

2. **Working Mechanisms**:
   a. Laws of Nature
   b. Constants of Physics
   c. Initial Conditions

3. **Objective** - The Sub-objective of fine-tuning a Universe is to achieve a stable Cosmological environment. And the main

objective of having a stable cosmological environment is to allow living beings to have an orderly, predictable, and reliable place to conduct their activities.

This objective assumes that the existence of DNA would be pointless if a stable cosmological environment were not available.

Humans depend on having a stable cosmological environment which gives them enough time to measure the overall human progress.

And to plan for future physical activities. Each person will participate in various activities which can bring about different outcomes. Those outcomes can range from dangerous to relaxing, disappointing to rewarding, tragic to happy, or simple to challenging.

From the variety of possible outcomes that could happen, each person's life journey would be different from other people's. One person may look at his life experiences as a waste of time for not having accomplished much. While another may see a more meaningful life experience for having accomplished many of his objectives. A person's perspective on life will depend on the person's openness to the maturing process that any person can only experience in a stable cosmological environment.

# Chapter 9: Human Aggression

Aggression - is usually associated with fighting, killing, and destruction. However, not all forms of fighting involve aggression. For example, fighting a condition to survive and fighting a war to survive are two different forms of fighting. On the one hand, fighting a condition so as to survive is a fight against the human condition. It is a non-aggressive struggle to secure food to temporarily escape the condition of death.

On the other hand, fighting a war to survive is usually a deadly struggle among nations to achieve some ideological supremacy. It is often an aggressive struggle to safeguard an ideology. The difference is that food is essential to existence, while ideology is not. Without access to food, humans could not exist. Without access to ideology, humans could still exist.

Whether humans are fighting a condition to survive or fighting a war to preserve an ideology, what is common between the two is that there is awareness of:

1. Condition
2. Ideology

Both condition and ideology are what humans are fighting for. With matter, however, there is no intelligence from which it can have an awareness of condition or ideology. Therefore, aggression or fighting is not only not relevant to matter, but it is also not a property of matter. Rather, aggression or fighting is a resistive, emotional mental activity response, which is relevant only to humans or animals who have access to intelligence. And can then apply or convert this emotional and mental response into physical reality in the form of Warfare or fighting.

# Chapter 10: Human Intelligence

"Intelligence - is defined as the mental capacity that allows us to reason, understand, solve problems, use abstract thinking and learn …" - Jaimar Tuarez.

Intelligence is the mental ability to do mental activities. Humans and animals have this mental or thinking ability whose origin we assume to be from another Intelligence or God. We are simply aware that we have this ability. Whether the mental Activity is simple or complex, what is observable is that our mental abilities control our physical activities.

Examples of simple mental activities:

1.  A 10-month-old child taking his/her first steps.
2.  A cat trying to escape a scientist's experimental enclosure.

Examples of complex mental activities:

1.  Solving the problem of anticipating the path of the hypersonic missile.
2.  Solving the problem of creating life.
3.  Solving the problem of living peacefully without having to rely on nuclear weapons.

Whether the Activity involves mostly mental such as learning Math, Language, or Music, or the activities involve both mental and physical such as playing Tennis or preventing Nuclear Destruction. There appears to be a pattern common to all activities: all activities, except sleeping, are problem-solving activities of how to get from point A to point C. Therefore, the mind only has to use one set of mental processing systems. No separate mental processing system for each Activity, with obstacles and the choice of solution as the only 2 things that would be different for each Activity.

There are at least 4 mental abilities which are foundational to all mental activities:

1. The ability to identify many types of information. Such as Principles, Coding, Fine-Tuning, Laws of Physics, Ideologies, Aggression, Gravitational Strength, Beneficial Traits, and Quality of Life.
2. The ability to understand the effects taking place when two or more pieces of information are interacting.
3. The ability to formulate or extract a solution or a response to counter the effect or to bring about the same effect as the one being caused by the interacting information.
4. The ability to apply or convert the conceptual solution or response into the physical reality.

Each mental ability works alongside a memory. Since these 4 foundational mental abilities work as one unit, then we must be dealing with a system that operates as a mental information processing mechanism we call intelligence.

The ability to apply is a combination of mental and physical abilities. On the mental side are awareness and understanding, and on the physical side are the human arms, legs, and bodies since humans and animals require certain physical products in order to function in physical reality. Physical products are first formed or extracted as an idea of a product, then converted into the physical reality as a physical product.

The difference between human and animal intelligence is that humans can express their ideas in written form, which helps to lessen human dependence on memory. Humans also have hands that can manipulate physical objects in all manner of orientations which helps with building tools as a way to test the applicability of new ideas in physical reality.

Animals, however, are not able to express their ideas in written form. They seem unable to see connections between different situations which may have similar patterns. They are then not able to or at least appear not able to form new ideas or new patterns from existing patterns.

Even if animals could form new ideas, they do not have hands with which to manipulate physical objects, which limits their abilities to use tools for specific use. As well as limit their abilities to create new tools for testing the applicability of their new ideas. On the contrary, Chimpanzees and Gorillas have hands, yet, they have not been able to create complex tools that are applicable to their basic functional needs.

What animals lack, or at least appear to lack, in their abilities to form new ideas, they make up with specialized abilities such as acute sense of hearing, smell, and vision or have better memory, stronger muscles, or faster running ability. In spite of being limited in intelligence, animals, in general, have a level of intelligence that allows them to adapt to their environment. They appear to have similar 4 foundational mental abilities that humans have.

Systems, principles, genetic coding, fine-tuning, sequencing, gravity, love aggression, nuclear threat, quality of life, laws of physics, music, and hyperrealistic drawings have relevance only to beings with intelligence who has the mental abilities to identify or have an awareness of the existence of such information.

If our 4 foundational mental abilities can identify and understand information such as Electromagnetic waves, DNA, Organ Systems, and Cosmological Fine-Tuning, which humans did not create. Then we can assume that another intelligence must have first identified, understood, formulated, and applied the physical creation or extraction of those physical items. These physical items could only have meaning to the one who created these items and to beings who could benefit from using these items.

# Chapter 11: Tools

A tool is practical, useful, and meaningful when it can be applied to human or animal physical and psychological functional needs.

The significance of a tool is that it is part of the Principle of Human Thinking. And also plays a big part in human adaptability in the environment. A tool bridges the gap between the non-material reality of the mind and the material reality of the physical body and the physical Universe.

A tool has several definitions:

1. It is used for a specific task or applied to a specific functional need.
2. It helps to compensate for certain limitations by:
    a. Helping the non-material mind to compensate for its inability to physically manipulate physical objects and therefore uses our physical bodies as its tool for manipulating other physical objects or other physical tools.
    b. Helping to compensate for certain physical limitations that our physical bodies may have. For example, we use telescopes because our eyes are not powerful enough to see objects at great distances.
1. It helps to transition non-material concepts into physical reality.
2. It helps our non-material minds to absorb or gain non-material knowledge. For example, Voyager I (tool) brings images of galaxies (knowledge) back to Earth. Scientists then view images of galaxies (non-material minds absorbing non-material knowledge).

Tools, however, do not just magically appear. It has to be identified by the mind. Fantasizing about being in a made-up reality, imagining new ideas, solving math problems mentally, or doing philosophy are things the mind may be good at doing. What the mind is not good at doing is manipulating physical objects.

The mind can think of a hammer, but it cannot physically lift a physical hammer. There are those who claim to have such ability.

However, most people never witness construction workers pound nails on a piece of lumber with a hammer using mind control. For most of us, our minds rely on something physical which can manipulate other physical objects; and that can only be our physical bodies.

The mind provides the intentions and the objectives, while the brain provides the electric impulses to all the muscle groups which move our arms, legs, and bodies. For example, the mind generates the intention for the hand to hold the hammer. The brain then sends electric impulses to the finger muscles, which cause the fingers to grab the hammer.

The mind then generates the intention for the arm with the hammer to be raised. The brain sends electric impulses, which cause the arm to move upward. Finally, the mind gives the intention for the arm to swing downward. The brain sends electric impulses to the raised arm, which causes the arm to swing downward. The hammer's momentum going down and hitting the nail drives the nail deeper into a piece of lumber.

If a tool is defined as anything that can be used for a purpose, then the mind becomes the initial tool. For now, we will consider the brain as the first tool and our physical bodies as the second tool, and the hammer as the third tool. And if we decide to create a table using the hammer and other tools, then that would become the fourth tool because we can then put things on top of the table, such as plates, books, computers, and cell phones. The combination of the mind, brain, and body has allowed humans to create a vast amount of tools.

How do tools fit in the human adaptation to the environment?

# Chapter 12: Principle of Human Thinking

The principle of human thinking consists of the 4 foundational mental abilities and 6 other abilities:

1. The mental ability to identify or have an awareness of information.
2. The mental ability to attach meaning to specific items of information to form knowledge.
3. The mental ability to store items of knowledge as information in a memory.
4. The mental ability to understand the effect an information is causing.

   Understanding is when what is being observed or what is being described is in agreement with certain items of knowledge stored in our memory.

5. The mental ability to recall certain items of knowledge from our memory.
6. The mental ability to identify a need relevant to the effect an information is causing.
7. The mental ability to formulate a response, solution, or objective relevant to the need.
8. The mental ability to identify the tools relevant to the objective.

   Tools either have to be:

   a. Created
   b. or Existing
9. The mental and physical abilities to apply the tools so as to convert, bridge, or transition the conceptual objective into the physical reality.

10. The mental and Physical abilities to control certain parameters to direct the relevant tools toward the objective.

How does the Principle of Human Thinking apply to everyday experience?

For example, a person still wrapped in his blanket slowly walks toward the front window of his house to take a look at the condition outside:

11. He identifies information that is presently in front of his house that was not there the night before.
12. He attempts to identify this information by searching through the various items of knowledge stored in his memory that matches the information outside.
13. He determines or identifies the information outside his house as snow with about 10 inches of accumulation.
14. He then attempts to understand the effect the information is causing and determines, based on his understanding, that the snow is covering his entire sidewalk.
15. He then attempts to identify the need relevant to the effect the information is causing and determines that people have the need to have safe, unobstructed access to the sidewalk.
16. He then attempts to formulate a response, solution, or objective based on that need. At this point, he has two options:
    a. Clear the snow later.
    b. Clear the snow now.
17. If he decides to wait until later, then that is the objective he has formed mentally. Which he will then convert or apply to his physical Activity when that time does arrive.
18. If he decides to clear the snow now, then that is the objective he has formed mentally, which he will then have to convert or apply to his physical Activity soon.

However, having formulated an objective will not automatically get us to our objective because our formulated objective exists in a non-material state of mind. While the objective we wish to accomplish is existing in the physical material state.

Therefore, we have a situation problem of how to get the non-material state which is point A, to the material state, which is point C. It is a gap that needs to be bridged.

1. To bridge that gap so that we can physically arrive at the physical objective state requires the mental ability to identify the tools relevant to that objective. These tools will either have to be:

   a. Created

   b. or already existing

   For example, a drill is a tool, but it is not relevant to the objective of clearing the snow.

   He determines or identifies the tools that he needs, which are: a snow shovel, Winter hat, Winter jacket, perhaps ski pants, gloves, and Winter Boots.

2. He can now apply, convert, or bridge the gap between the conceptual objective in the non-material state and the physical objective in the material state. By using his hands as the 2nd tool (the brain being the 1st tool in the chain of tools) to manipulate the 3rd tool, which is the shovel which he can use to start clearing the snow.

3. He has to control certain parameters to make sure that the tools are being directed toward the objective. Meaning that the snow shovel can be monitored, observed, or verified as being used in shovelling the snow on the sidewalk, not the snow on the street.

The principle of Human Thinking also applies to the manufacture of products. For example, when attempting to create a product such as a car, it starts with a conceptual objective of designing a car in the non-material state, which has to be converted into the physical, material state. Converting the concept of a car into the physical is simply a matter of applying the relevant physical tools.

The first conversion is with the physical medium of ink and paper. Using our hands to hold the pen, we would then guide the pen on the paper to draw the image of the car as well as to draw the image of the instructions in letters, characters, numbers, and other symbols on how to assemble the car. Other conversions to the physical would then be based on those instructions.

The number of conversions to the physical will depend on the number of parts. However, each conversion to the physical, as far as we know, will have 3 stages.

> **Stage 1** is the starting stage, where all the identifying, understanding, and objective formulating are happening in the non-material realm of the mind.
>
> **Stage 2** is the transitional stage, where tools are being applied to convert the conceptual mental activities into material, physical activities.
>
> **Stage 3** is the destination stage, where objectives become completed. Either as a physical product or as a physical activity that improves someone's quality of life. Such as Clearing the snow on the sidewalk or Agreeing to peaceful coexistence.

Physical activities or body movements would be different for different activities, but the mental procedures remain the same.

For example, playing the piano, operating an airplane, or constructing a bridge would require different body movements and different sets of knowledge. However, the mental procedure of identifying, understanding, and formulating would be the same for all activities.

Because the principle of human thinking is broken down into a step-by-step procedure, this then makes the principle of human thinking a repeatable, testable, demonstrable, accountable intelligence-based information processing system and also an intentional, objective-based interactive mechanism.

Obviously, this is simply to show the general breakdown of the principle of human thinking. It is not, however, advisable to start breaking down our daily activities into steps as this will only have the effect of slowing our reaction time because we would then be confusing the mind by re-introducing something the mind had already learned at an early age.

The mind has become very good at following the thinking procedure that the process has become automatic. The side effect is that everyone, meaning every mind, has forgotten how the procedure of thinking is done. The mind is simply doing the thinking procedure without remembering how it is applying the procedure. It is like learning how to speak a particular language without ever learning how to read a word from that language.

# Chapter 13: Newton's Law of Motion

Newton's law of motion helps to confirm that the physics of motion also applies to the formation of life, fine-tuning, genetic coding, the Big Bang, and hyperrealistic drawing. Since the Activity of any type of formation involves direction of movement or motion; therefore, molecules or particles cannot move in any direction on their own. Meaning that matter cannot move toward the formation of galaxies, the formation of hyperrealistic drawings, or the formation of life without violating Newton's law of motion.

"A body at rest will remain at rest…unless it is acted upon by an external force…" - *Isaac Newton.*

Matter has no intelligence to have any mental ability to identify itself, direction, or time as existing information. Therefore, it has no concept of its directional condition today nor how its directional condition in the future will be. It has no mental ability to formulate the objective of changing its direction of motion today in the hope that its motion will have a better direction in the future or hope that its direction is heading toward creating life, calibrating gravity, encoding DNA, or drawing hyperrealistic images.

# Chapter 14: Mind Body Duality

" . . . how a thing moves depends solely on (i) how much it is pushed, (ii) the manner in which it is pushed, . . . The first two of those require contact . . . the third requires that the causally active thing be extended, Your notion of the soul entirely excludes extension, and it appears to me that an immaterial thing can't possibly touch anything else . . . "

Princess Elisabeth of Bohemia

Scientists and Philosophers today are committing the same mistake that Princess Elisabeth made in 1643, which is the failure to understand the mind body duality. As humans, we have five senses which are instruments or tools for detecting information. However, our five senses can only detect physical information.

The problem arises when scientists, philosophers, as did Princess Elisabeth are attempting to apply their five physical senses: sight, sound, smell, taste, and touch to detect that which is non-physical or non-material. That would be like trying to insert a compact disc into an SD card slot.

This could never work as the two medium formats are incompatible. A conversion has to happen for the two opposing formats to work. In the same way that a digital music file cannot be listened to unless it is first converted into analog. This digital music file can be expressed as recognizable sound by using a speaker. The vibrating motion of the speaker cone will produce sound waves that the human ear can hear, which the mind can then translate as music.

Having limited knowledge is a characteristic of being human. But being limited in knowledge does not mean incapable of adding new information to our knowledge. However, there is a limit to how much we can add to our knowledge.

For example, Bonobo Apes have been shown to have the ability to create basic language by stringing together different images that have meanings attached to those images. However, since Bonobo Apes do not appear to have figured out the idea of using tools to bridge the conceptual into the physical, this then limits what they can add to their knowledge. For example, the knowledge of creating complex tools such as Airplanes, Electron Microscope, and Cell Phones which would most likely remain elusive to Ape intelligence.

In a similar fashion, since the tool for detecting non-material information is not available to us, this then limits what we can add to our knowledge. For example, the knowledge of understanding the mechanism behind the conversion of non material information into some physical medium. Which should add to the knowledge of understanding what material components must be combined, separated, or converted that could cause matter to come to life, create consciousness, or make infomation encodable in DNA. But, instead these would most likely remain elusive to human intelligence.

Obviously, there is someone or something who has knowledge on how the contrasting substances of mind and body can be converted into a functioning mechanism. Or we would not be having this experience where our mental  activities have power to direct our physical activities.

# Chapter 15: Mind-Matter Interaction

How does the mind interact with the Physical World? Many have a hard time envisioning a non-material mind having the ability to move solid objects. Let us examine how the human mind can move objects. The human mind interacts with the physical world by converting conceptual thoughts, ideas, and objectives into the physical medium of electric impulses.

MRI machines might be able to show electric impulses firing within certain brain sections, but MRI machines cannot tell scientists whether those electric impulses carry messages. For example, the messages, Bring flowers to Mom, and There is an accident up ahead. To humans, these two sentences are different just from the understood meaning each is expressing. However, to MRI machines, the two are interpreted as simply glowing lights being given off by some section of the brain.

We know that if we wish to grab, for example, a Rubik's cube, our hands somehow will follow what we wish them to do. We also know that what moves our muscles are electric impulses. From the moment we make the intention to move our hand toward the Rubik's cube to the moment electric impulses are causing our finger muscles to grab the Rubik's cube is a process that happens instantaneously.

Such that the intent to move the hand and finger muscles by applying pressure on the Rubik's cube to lift it appear to happen at the same time. However, somewhere between the short period of synchronized movement, our non-material thought of picking up the Rubik's cube got converted into the physical medium of electric impulses.

Did our thought of picking up the Rubik's cube become imprinted within the electrons of the electric impulses? We can only assume that electric impulses are there only to cause certain muscle groups to contract. What might be happening Is there a message converter that can read non-material messages then converts it into some mechanical energy that can trigger the release of electric impulses.

This converter would be like a Digital to Analog converter. But instead of using computer chips, it uses the physical medium of enzymes. Similar to the enzymes in DNA. But instead of translating messages into creating proteins, it identifies which messages carry the intent to move certain muscles, as well as identifies which muscle groups the message is intending to move. Then it sends the electric impulses to the intended destinations.

Obviously, not all messages require sending electric impulses to move some muscles. Because not all thought processes require our hands to pick up an object or require our feet to take a step forward. While the non-material message converter is an assumption; its operation, where an enzyme triggers the release of electric impulses is for now plain speculation.

Because, how our non-material thought of picking up a Rubik's cube is suddenly replaced by the physical medium of electric impulses is a knowledge that is not available to us at this stage. Or not detectable by our physical senses, nor detectable by MRI machines. We may not see how our thoughts are being replaced with electric impulses, but we can feel or experience the effects of those conversions by being able to move our hands.

There is of course a second phase to this mind-to-matter conversion. Which focuses on the physical objects the mind is able to manipulate, and the physical products the mind is able to create just by being able to control our hand movements. In this second phase, we are still getting instructions from the mind, and electric impulses are still being released. It is simply a continuation of the first phase.

The second phase begins with the hand holding, for example, a hammer:

1. By being able to control how the hand moves, the mind gains the ability not only to hold tools such as a hammer but also the ability to manipulate those tools

2. By being able to manipulate other tools, the mind gains the ability to create more tools. Such as a table, chair, house, computer, or a cell phone

Therefore, the mind controls the movements of our hands to hold a tool. Then manipulating that tool with our hands to create more tools is how the mind fulfills its functional needs, as well as how it interacts with the physical reality. This is the mind and matter interaction on the human side. Let us now turn to the mind and matter interaction on the supernatural side.

If a supernatural being could interact with the physical reality, it would most likely do so through Mathematics. Numbers being conceptual means it has the same non-material property as the assumed supernatural being. With its neutral property, numbers can be applied to different physical items or physical features. Such as level of energy, strength of gravity, and speed of light.

The application of numbers means values can be set more accurately. Having a number that identifies the exact location where a value is being intended to be placed is a more superior method to Nature's Wheel of Fortune-Tuning method of randomly selecting a location. To want to have an accurate reading is a desired objective. Formulation of objectives and conception of numbers are not properties of matter, but activities of the mind.

How does Nature fine-tune a Universe without using numbers? Lacking Intelligence, Nature has no mental ability to conceptualize numbers. Without any concept of numbers, what identifying system could Nature use to accurately set:

a) The strength of Nuclear Force

b) The strength of gravity

c) The speed of light

d) The expansion rates

With values of Nuclear Force, Strength of Gravity, and Speed of Light not clearly defined, how does Nature establish any mathematical equation for each parameter? With each parameter having no established mathematical equation, how does Nature establish the Laws of Physics? If parameters have inconsistent values and cannot be used to establish the Laws of Physics, what mathematical equations does Nature combine to arrive at a finely-tuned Universe?

" The Mandelbrot Set does not occur in Nature. However, the mathematical patterns that produce the Mandelbrot set do . . . "

Fibonacci Numbers and the Mandelbrot Set

$Zn+1=Zn^2+C$ is the Mandelbrot equation discovered by Benoit Mandelbrot. When fed into a computer, this equation will produce a 2 dimensional printout of repeating geometric patterns. According to many scientists, similar patterns can be found in Nature. The examples they give are:

Sea Shells, Ferns, Sunflowers, Pinecones, Hurricanes, and Spiral Galaxies. However, these examples are misleading. Because Romanesco Broccoli, Pinecones, and Sunflowers have DNA, which means they carry genetic codes. However, coding is not a property of matter, but an activity of the mind.

How about Hurricanes and Spiral Galaxies? Both have been shown to fit the shape pattern of the Fibonacci Sequence. This is a type of pattern that can be found in Nature including Pinecones, Sunflowers, and Pineapples. The Fibonacci Sequence is named after Leonardo Fibonacci who discovered it. Both the Fibonacci Sequence and the Mandelbrot Set are number-based pattern-generating systems. Neither of which is a property of matter, but an activity of the mind.

How conceptual numbers from a supernatural being can be converted into something applicable in the physical reality is a knowledge that is not available to us, or not detectable by our physical senses. We may not see how this conversion is being done, but we can feel or experience where in Nature these numbers are being applied. For example, objects still fall to the ground; the same ground that keeps us attached to this planet. Patterns in Pinecones, Sunflowers, and Hurricanes still follow the Fibonacci Sequence.

The calibration of the strength of nuclear force, the strength of gravity, and the speed of light. The spiraling pattern of the sunflower seeds, the spiraling shape of the nautilus shell, and the spiraling formation of a galaxy. These are trails of activities or interactions within the physical reality. Which either uses a mathematical formulaic equation or mathematical pattern-generating sequence. Whose conceptual origin point to a non-material, highly intelligent mind.

# Conclusion

Wind can extract order in the form of Sand dunes, but it is not Nature showing its ability to extract order out of chaos. The wind is simply obeying a property, and the property of matter is basically all the physical ability that Nature has as its main order-extracting mechanism. The problem with the property of matter is that it carries only one physical function. Therefore, it can only give the wind the ability to do one physical function. It cannot give the wind the ability to do multiple physical functions or complex physical processes.

The reason matter cannot do complex physical processes is that it is missing several items:

1. Physical Abilities: Organize, Control, Manipulate
2. Mental Abilities: Identify, Understand, and Formulate
3. Applicable Principles: Step-by-step procedures.

Without physical abilities, matter cannot physically organize, control, or manipulate sand into cinder blocks, manipulate colors into hyperrealistic images, or manipulate amino acids into proteins.

Without mental abilities, matter cannot mentally identify itself, condition, and time as existing information. Therefore, it cannot have any concept of its condition today or its condition in the future. Therefore, matter could not formulate the objective of holding on to some features in the hope that it could potentially add complexity to improve its future condition.

Without principles, matter does not have a step-by-step procedure on how its physical abilities could extract cinder blocks out of the sand, extract hyperrealistic images out of color materials, or extract proteins out of amino acids.

The scientists' concept of evolution and the supposed physical abilities of Nature rest on the 4 fundamental forces of Nature and the chemical bonding of particles. However, these forces of Nature, including the properties of matter, cannot be converted into any applicable principles.

Unable to gain any useful principles from the Natural forces, Scientists are then left to rely on their imaginations as their guiding principles on evolution. But whose premise is based on their beliefs that matter and that Natural forces are able to do complex physical processes. Or are able to mutate simple organisms into complex organisms by way of matter's supposed ability to encode DNA.

Unfortunately, fundamental forces, properties of matter, and accidental chemical reactions could only reach the lower stages of order formations or order extractions. It could never reach the higher stages of order formations or the stage where only intelligence has the ability to arrange formations with meaningful complexity.

# References

1.  A Universe Not Made For Us.
    YouTube - Carl Sagan

2.  Gravity
    newscientist.com/definition/gravity/ - Richard Welob

3.  The Weak Force
    webs.morningside.edu/slaven/physics/micro/micro7.html

4.  Strong Nuclear Force
    What is Nuclear Force Residual Strong Force Definition 2019-
    05-22 - Nick Connor
    periodic-table.org/what-is-nuclear-force residual-strong-force-
    definition

5.  Electromagnetic Force
    Astrophysics ecuip-lib-uchicago-edu/multiwavelength.
    astronomy/astrophysics 05-html

6.  Chemical Bonds
    Anatomy & Physiology
    Lindsay M. Biga, Sierra Dawson, Amy Kaufmann, Mike Le
    Master, Phillip Mater n, Kati Morrisson Graham, Devon Quick
    & Jon Runyeon
    open.oregonstate.educationaandp/chapter/2.2chemical bonds

7.  Surprising Origin of Evolutionary Complexity Scientific
    American - Aug 2013
    scientificamerican.com/article/the-surprising.origin.
    of.evolutionary-complexity

8.  What is DNA? Medline Plus National Library of Medicine
    medlineplus.gov/genetic/understanding/basics/dna/

9.  System Meaning
    System Meaning/best 35 Definition of system
    yourdictionary.com/system

10. The Fine-Tuning of the Universe      drcraigvideos      YouTube
    2017

11. What's the Fluid Intelligence Definition in Psychology? Jaimar
    Tuarez - Jun 20, 2022
    neurotray.com/whats-the-fluid-intelligence-definition-in-
    psychology/

12. What is Classical Mechanics? Robert Coolman    Sept 12, 2014 -
    Newton's Law of Motion      Isaac Newton
    livescience.com/47814-classical-mechanics.html

13. 13-year-old art prodigy
    Isabellaclever.com

        Elisabeth writes on 6.V. 1643: Correspondence between
        Descartes and Princess Elisabeth Copyright Jonathan
        Bennet 2017 earlymoderntext.com/assets/pdfs/descartes
        1643_ 1,pdf

        Wheel Of Fortune is owned by Sony Pictures Television
        Studios

        A Sony Pictures Entertainment Company

        Fibonacci  Numbers  and  the  Mandelbrot  Set
        fractalfoundation.org

        Rubik's Cube is owned by Rubik's Brand Ltd

www.ingramcontent.com/pod-product-compliance
Lightning Source LLC
Chambersburg PA
CBHW051246120626
46547CB00014B/1818